JUL 0 3 2016

DATE DUE

FEB 2 0 1999	JUL 2 8 2005	
APR 1 2 1999	DEC 0 6 2005	JUL 0 6 2013
JUL 1 8 1999	MAR 2 1 2007 MAY 2 3 2007	FEB 2 8 2015
SEP 0 8 1999	JUN 2 6 2007	
JUN 2 1 2001	AUG 0 2 2011	
JUL 2 3 2001	MAY 0 7 2015 APR 3 0 2016	
JUN 0 5 2002	AUG 0 9 2016	
Nov 16 2002 MAY 2 5 2003		
SEP 1 6 2003		

FLYTRAP

pitcher plant, Borneo

LIVING THINGS

FLYTRAP

Rebecca Stefoff

BENCHMARK BOOKS

MARSHALL CAVENDISH
NEW YORK

Benchmark Books
Marshall Cavendish Corporation
99 White Plains Road
Tarrytown, New York 10591

Library of Congress Cataloging in Publication Data
Stefoff, Rebecca.
Flytrap / by Rebecca Stefoff.
p.   cm. — (Living things)
Includes index.
Summary: Photographs and text describe the flytraps, sundews, butterworts, and
other plants that consume insects, worms, and even small frogs.
ISBN 0-7614-0445-7 (lib. bdg.)
1. Carnivorous plants—Juvenile literature. 2. Venus's flytrap—Juvenile literature.
[1. Carnivorous plants. 2. Venus's flytrap.] I. Title. II. Series: Stefoff, Rebecca
Living things.
QK917.S74   1999   583'.75—dc   97-1192   CIP   AC

Photo research by Ellen Barrett Dudley

Cover photo: *The National Audubon Society Collection/Photo Researchers, Inc.*,
J.H. Robinson

The photographs in this book are used by permission and through the courtesy of:
The National Audubon Society Collection/Photo Researchers, Inc.: Simon D.
Pollard, 2, 14 (left); Dan Suzio, 6-7; Ray Coleman, 7; Dr. W.M. Harlow, 8 (bottom);
Nuridsany et Perennou, 9, 10, 23, 25, 27; Farrell Grehan, 12; S. McKeever, 12-13,
16; Miiton J. Heiberg, 13; Dr. Paul A. Zahl, 14 (right), 15 (right); J.H. Robinson, 15
(left); Fletcher & Baylis, 17 (left); L. West, 17 (right); Dr. Jeremy Burgess/Science
Photo Library, 19; Jerome Wexler, 20; D. Nitidula, 20 (inset); Jeff Lepore, 24, 32.
Earth Scenes: Zig Leszczynski, 8 (top); Alastair Shay, 21. *Peter Arnold, Inc.*: Walter
H. Hodge, 11; Ed Reschke, 18, 22; Fred Bavendam, 26.

Printed in Hong Kong

1   3   5   6   4   2

For all my vegetarian friends

Venus flytrap

flytrap's leaf

This is a plant with a secret. It's a meat eater. Each one of its leaves is like a hungry mouth.

When a fly wanders onto the leaf, the two sides of
the leaf snap shut. The fly is trapped inside. That's
how this plant got its name: the flytrap.

The flytrap's leaves give off chemicals that turn the
fly's insides into juice. This juice soaks into the leaves
and helps feed the plant. After a few days the leaf
opens. By now the fly is just a dried-out shell.

How does the leaf know when to snap shut? Look
closely at the open red trap with the dried-up fly in
it. Do you see a few tiny pointed hairs inside the
trap? When something brushes against these, the
trap closes.

The flytrap catches more than just flies. It can trap and eat a katydid, a bee, or even a small frog. It's an amazing plant—but it isn't the only plant that eats bugs and animals.

pitcher plants, Georgia

sundews

All of these plants are meat eaters, too. They trap insects, worms, spiders, and other creatures. Like the flytrap, these plants make chemicals that turn their prey into liquid plant food.

There are about 450 kinds of meat-eating plants in the world. They have several different ways of trapping bugs and small animals.

thread-leafed sundew, New Jersey

These are pitcher plants. Their leaves look like pitchers. The pitchers come in many shapes and sizes, but all of them hold liquid. Some pitchers are full of rainwater. Other plants make sweet-smelling juice to fill the pitchers.

pitcher plants, Borneo

Thirsty insects or small animals crawl into the pitchers to drink the water or juice. Once they're inside, they can't get out. Sharp, downward-pointing hairs keep them from climbing back out of the pitcher. They drown and become food for the pitcher plant.

trumpet pitcher plants

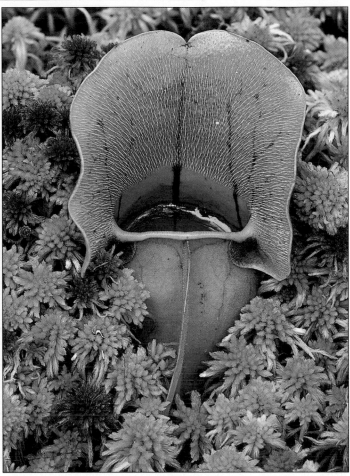

Pitcher plants grow in North and South America, Asia, and Australia. The smallest pitchers are less than two inches (5 cm) tall. The largest are about four feet (120 cm) tall, with thick, slippery walls. These giant pitchers can trap prey as big as a rat. Sometimes frogs or salamanders live inside the pitchers, feeding on insects and other prey trapped there.

round-leafed sundew

Sundews use sticky drops, not slippery pits, to trap their prey. Special hairs on the sundew's leaves give off a clear liquid. The liquid sparkles in the sun. It looks like water, or dew. But when thirsty insects land on the sundew, they find that the liquid isn't water at all. It's more like glue.

Once an insect is stuck to the leaf, the leaf starts to curl. It wraps itself around the helpless bug. Chemicals from inside the leaf begin the slow job of drying out the prey.

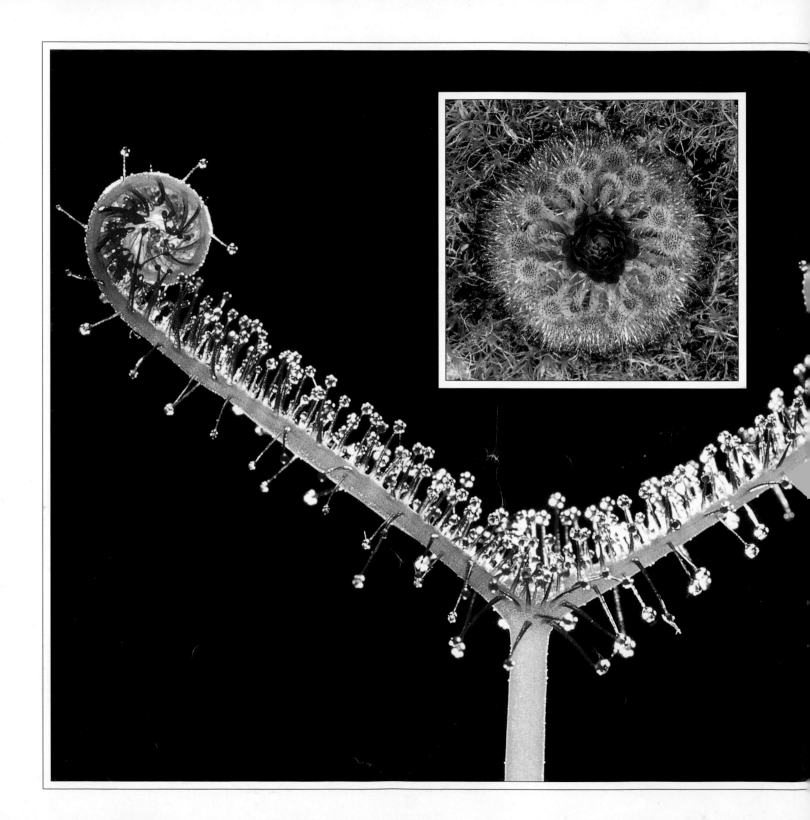

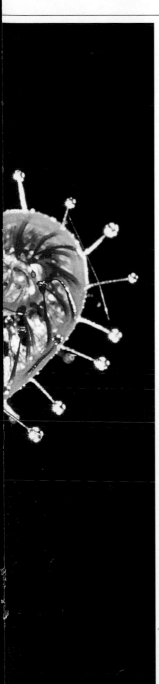

There are more than a hundred kinds of sundews in the world. Many of them grow only in Australia, but others are common in parts of Europe, Russia, Canada, and the United States. Their leaves and flowers come in many different shapes, but all of them sparkle with bright drops of deadly "dew."

butterwort, North Carolina

This is a butterwort plant. Its leaves are covered with thousands of very tiny hairs. When an insect touches the hairs, small openings in the leaf pour out a sticky jelly that looks like butter—that's how the butterwort got its name. The jelly traps the insect while the edges of the leaf curl inward around it. This small fly landed on a butterwort leaf. Now it discovers that it can't get its legs out of the sticky mess.

fruitfly caught on butterwort leaf

bladderworts blooming among pond cypress trees

The last kind of meat-eating plant is the bladder-wort. The small yellow flowers blooming just above the water in this swamp are bladderwort flowers. But the bladderwort's trap is under the water, where its prey lives.

Its long and branching leaves are covered with little sacs. Tiny water creatures, some smaller than the dot on this i, swim past the leaves. When one of these creatures brushes against hairs on one of the sacs, the sac sucks in water— and it sucks in the creature, too.

bladderwort's underwater traps

northern pitcher plants, New Hampshire

Many kinds of flytraps, pitcher plants, sundews, and butterworts like sunshine and moist soil. They live in open, wet fields called bogs. If you want to see one of these marvels of the plant world in action, look for a sunny spot that's damp underfoot.

A QUICK LOOK AT THE FLYTRAP

Flytraps, pitcher plants, and the other plants in this book are sometimes called insect-eating plants. A better name, though, is carnivorous, or meat-eating plants, because these plants eat spiders, worms, snails, frogs, and other small creatures in addition to insects. Carnivorous plants grow where the soil is poor. Eating animals is the plants' way of getting nitrogen and other nutrients that they can't get from the soil. They use three types of traps: pitfall traps (pitcher plants); sticky traps (sundews and butterworts); and spring traps (flytraps and bladderworts). About 450 species, or kinds, of carnivorous plants are known.

Here are six kinds of carnivorous plants, along with their scientific names and a few key facts.

COMMON OR NORTHERN PITCHER PLANT
Sarracenia purpurea (seh ruh SAYN yuh pur puh RAY uh)
One of most widely distributed pitcher plants. Found in bogs or wet, sandy meadows throughout eastern North America, north to Arctic Circle and west to Great Plains.

VENUS FLYTRAP
Dionaea muscipula
(dee on NEE yuh myew skih PYEW luh)
Found only in North and South Carolina, in moist areas near bogs and pine forests. Each leaf is a trap that can close in one-tenth of a second.

28

PURPLE-FLOWERED BUTTERWORT
Pinguicula group (30 or so species)
(pin gwi KYEW luh)
Found in cool climates in United States, Canada, Europe, and Russia. Sometimes called bog violets because many species have purple flowers that rise above leaves on slender stems.

LONG-LEAFED SUNDEW
Drosera group   (DROSS er uh)
One of many species of long-leafed sundew, most of which are native to southwestern Australia. Uses a combination of sticky fluid and curling leaves to trap and hold prey.

COBRA PLANT
Darlingtonia californica
(dar ling TONE yuh cah lee FOR ni kuh)
Species of pitcher plant found only in Oregon and California. Hood, or flap, over pitcher resembles the head of the hooded cobra snake. Pitchers can be two feet (61cm) tall.

29

GIANT PITCHER PLANT

Nepenthes rajah

(neh PEN thees RAH jah)
World's largest pitcher plant.
Has a climbing stem with leaves
that form pitchers. A pitcher
may be four feet (122 cm) tall.
Can capture prey as large as a
rat. Found only on the tropical
island of Borneo, in environ-
ments ranging from edge of
beach to mountain highlands.

Taking Care of the Carnivorous Plant

Many carnivorous plants grow only in special places like bogs, wetlands, and
sandy meadows. We must be careful to leave enough of these places undisturbed
so that the plants can survive. Another threat to some species of carnivorous
plants is collectors. Lots of people enjoy raising flytraps and other carnivorous
plants at home, but so many people take plants from some wild places that the
plants are in danger of being wiped out. If you want to raise a carnivorous plant,
buy it from a nursery that raises plants and does not take them from the wild.

Find Out More

Clyne, Densey. *Plants of Prey*. Chicago: Independent Publishers Group, 1993.

Gentle, Victor. *Bladderworts: Trapdoors to Oblivion, Pitcher Plants: Slippery Pits of No Escape, Sundews: A Sweet and Sticky Death*, and *Venus Fly Traps and Waterwheels: Spring Traps of the Plant World*. Milwaukee: Gareth Stevens, 1996.

Jenkins, Martin. *Fly Traps!* Cambridge, Mass.: Candlewick Press, 1996.

Kite, Patricia. *Insect-Eating Plants*. Brookfield, Conn.: Millbrook Press, 1995.

Nielsen, Nancy. *Carnivorous Plants*. New York: Franklin Watts, 1992.

Wexler, Jerome. *Sundew Stranglers: Plants that Eat Insects*. New York: Dutton Children's Books, 1995.

Index

Rebecca Stefoff has published many books for young readers. Science and environmental issues are among her favorite subjects. She lives in Oregon and enjoys observing the natural world while hiking, camping, and scuba diving.

white-topped pitcher plant, Alabama